Bibliographic information published by the German National Library:

The German National Library lists this publication in the National Bibliography; detailed bibliographic data are available on the Internet at http://dnb.dnb.de .

Imprint:

Copyright © 2019 GRIN Verlag
Print and binding: Books on Demand GmbH, Norderstedt Germany
ISBN: 9783346023452

This book at GRIN:

https://www.grin.com/document/500400

Alexander Markus Rehm

Why Cattle love the Saxophone sound but ignore Crane calls?

Analysis of sounds from Cattle, Cranes, Tenor Saxophone and Kulningar

GRIN Verlag

GRIN - Your knowledge has value

Since its foundation in 1998, GRIN has specialized in publishing academic texts by students, college teachers and other academics as e-book and printed book. The website www.grin.com is an ideal platform for presenting term papers, final papers, scientific essays, dissertations and specialist books.

Visit us on the internet:

http://www.grin.com/

http://www.facebook.com/grincom

http://www.twitter.com/grin_com

Why Cattle love the Saxophone sound but ignore Crane calls?

Analysis of sounds from Cattle, Cranes, Tenor Saxophone and Kulningar

Author: Alexander Rehm

Summary:

The observation that Cattle which are spread over a wider area gather around a Saxophone player but totally ignore Crane calls gave the impetus for a detailed investigation of the "timbre" of both sounds as a possible explanation for the observed effect. Frequency- and Formant-spectra of four different sounds have been analysed: Cattle-calls (cowing), Kulning (an ancient herding call of Swedish women), Crane-calls and Saxophone play (by a professional player). Kulning-calls and the Saxophone-sound exhibit Formant-spectra with several Formant-bands of high intensity in the range of 1.000-9.000Hz which are nearly completely missing in Formant-spectra of Crane calls. Therefore, it is concluded that this richness of sound (=timbre) of Saxophone and Kulning generate the "attraction" of these sounds for Cattle whereas Crane-calls which miss this Formant-signals nearly completely cannot gain the interest of Cattle.

Introduction:

The following observations and published data generated the impetus for this study:

1) It can be observed that Cattle are often cowing for 2 reasons: a) a mother is calling her child and b) if a regular procedure, the Cattle are used to, has been changed (e.g. the milking time is delayed). In both cases the Cattle is cowing to be heard by a certain species of "their herd" (either the child or the "herd-member" doing the milking) – it is a call for action.

2) The ancient herding call "Kulning" from Sweden is very efficient to attract the attention of Cattle and to gather them around the "Kulningar" (see Ref. 2).

3) Calls of Cranes during the spring and summer period are nearly completely ignored by Swedish Cattle. Even young Cattle which have not heard Crane calls before do not show any significant reaction (e.g. moving up their head or turning their head in the direction of the call) during or right after a Crane call.

4) Playing the tenor Saxophone very often generates the following reaction within a herd of Cattle: With the first sound of the Saxophone the Cattle move up their head, look in the direction of the Sax player and start to move in the direction of the player and will gather around him in a distance of 4-10 meters. This "experiment" has been successfully repeated several times with different herds of Cattle in Smaland/Sweden.

These data raise the following questions:

Why do Cattle ignore Crane calls although these calls have a similar basic frequency around 800-1000Hz as "Kulning" which obviously gets their attention?

Why is a Saxophone sound so attractive for Cattle although this sound differs significantly from the sound of "Kulning" or the sound of cowing?

As Cattle are in general curious the simplest explanation would be that they show the above described behavior as a reaction on an unexpected noise or sound. But this would not explain the massive difference between the "attractor-effect" of Kulning or Saxophone sound and the "ignoring behavior" upon Crane calls. So it might be the character of the sound itself which may makes the difference. To understand more about these phenomena sounds of cowing, Kulning, Crane-calls and Saxophone play have been analysed and compared in order to find arguments why Cattle show different reactions to these sounds. In this publication it is also considered that Cattle have a significantly different "hearing curve" (see "Figure 1" extracted from. Ref. 1) compared to humans which seems to be of importance to understand the above described phenomena.

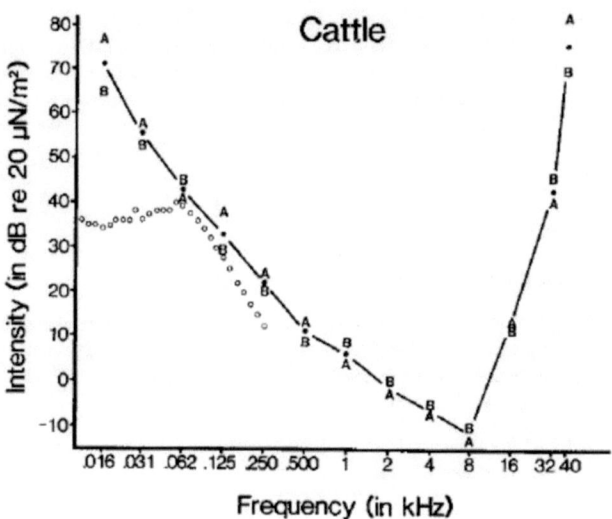

Figure 1: Hearing curve of Cattle (individuals "A" & "B"; open circles = background noise) extracted from Ref.1.

Material and Methods:

The sounds analysed in this publication are from the following sources:

a) Crane calls are downloads of sound files from youtube (Ref. 3; 4)

b) Kulning-sounds are downloads of sound files from youtube (Ref. 2; 5)

c) Cattle cowing calls are a download of a sound file from youtube: (Ref. 6) and recordings from Cattle in Smaland/Sweden done with a mobile recording equipment.

d) Saxophone sounds have been recorded according to Ref. 7

2

The sound files have been analysed and processed using the software "Praat" as described in Ref. 7 and Formant spectra have been generated according to Ref. 8.

Results:

All 4 sounds under investigation (Crane calls, Cowing, Kulning, Saxophone play) show two common characteristics:

1) The sounds itself show a variation in the basic frequency within the time period of 2-4 seconds which can be defined as "basic melody pattern of the sound".

2) The basic melody pattern of the four sounds are repeated several times with either slight variations (Crane calls and Cowing) or larger variations (Kulning, Saxophone).

So it can be assumed (although this has not been investigated or proven in this study) that in general a repetition of a short "basic melody pattern" with some variations may have an "attractor function" for Cattle. In other words: a variation and repetition of a basic melody pattern lasting some seconds might be a prerequisite to gain the attention of Cattle.

Cattle call:

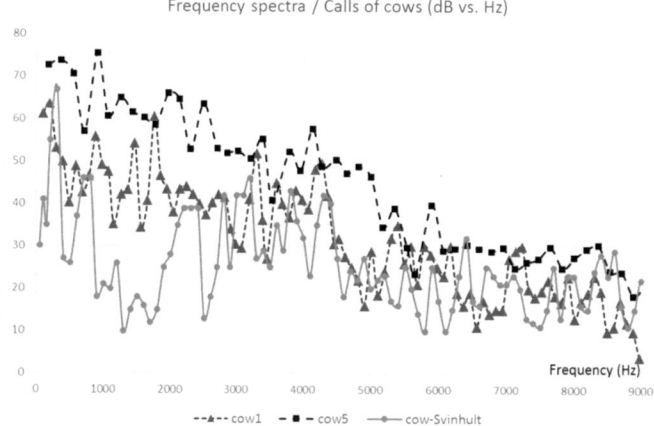

Figure 2: Frequency spectra of different cow calls derived from a phase of a relative stable basic frequency and a stable overall sound intensity. Cow1 is calling with a basic frequency of 96Hz, cow5 with a basic frequency of 176Hz whereas cow-Svinhult is not calling with a clear defined basic frequency. (Y-axis: rel. dB / X-axis: Hz)

It can be further assumed that the most familiar sound for Cattle is the call (cowing) of another Cattle. Therefore the calls of Cattle should be analysed in order to get an idea of the typical characteristics of a sound which is familiar to Cattle. Figures 2 and 3 show Frequency spectra and the respective Formant spectra (derived from frequency spectra according Ref. 8) of typical Cattle calls.

3

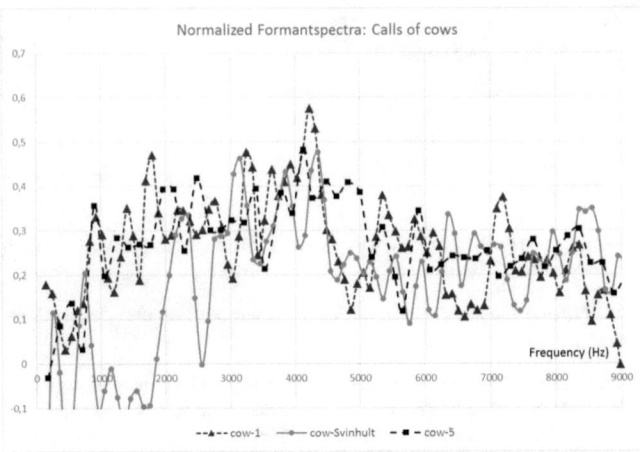

Figure 3: Normalized Formant spectra derived from the Frequency spectra in Figure 2 according to the method described in Ref. 8. (Y-axis: rel. dB / X-axis: Hz)

The following differences and common characteristics of the calls are obvious in the Formant spectra:

a) Although the calls show significant differences in the basic frequency of the call, all calls exhibit a strong Formant-signal in the range between 3.000-4.500 Hz and Formant bands in the range of 5.000-9.000 Hz with lower intensity.

b) Calls of cow1 and cow5 are very similar in Formant expression in the range between 1.000-3.000Hz whereas Formant signals of cow-Svinhult are nearly missing in this frequency-range.

Kulning:

In northern Europe and especially in Sweden "Kulning" is a traditional herding call from women to gather Cattle which are spread over a larger area. The "Kulning-calls" are very effective to gather Cattle so it can be assumed that these calls are easily recognized by Cattle and generate some attraction for Cattle, so they gather around the "Kulningar". Inspired by existing research on "Kulning" (see Ref. 9 & 10) Formant-analysis of various Kulning-calls have been performed. Typical Formant-spectra of Kulning-calls are exhibit in Figure 4. Formant spectra of two different phases of a Kulning call (which in total can last several seconds) which have different basic frequency are shown. Beside prominent Formant-signals in the range of 3.000-5.000Hz additional Formant-bands with strong intensity in the range of 7.000-9.000 Hz can easily be recognized. Compared to Cattle-calls (=cowing) Kulning shows much stronger Formant-signals in the frequency range >7.000Hz (see Figure 5) which this is the frequency range where the hearing curve of Cattle (see Figure 1) shows a minimum which means that Cattle have a high hearing-capability in this range.

4

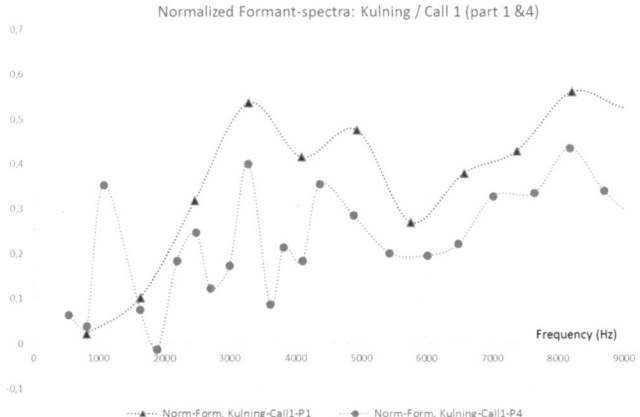

Figure 4: Normalized Formant-spectra of two parts of a Kulning-call from Ref. 2. Part 1 has a duration of 1,4 seconds and shows a harmonic structure with basic frequency of 819Hz. Part 4 has a duration of 0,9 seconds with a basic frequency at 543Hz and a complex overtone structure.

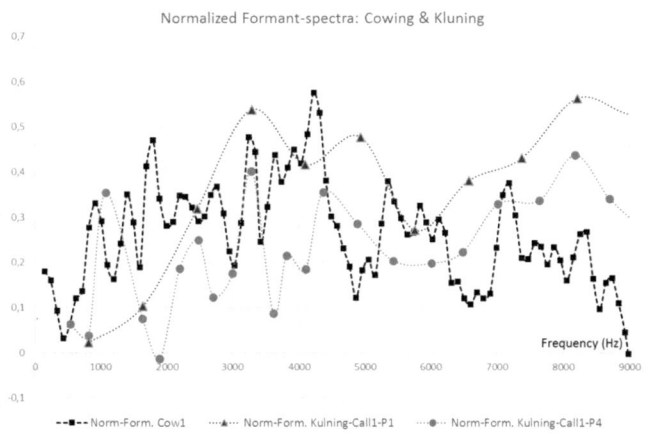

Figure 5: Comparison of the normalized Formant-spectra of Kulning from Fig. 4 and a Formant-spectrum of a cow-call from Fig. 3 (squares).

Saxophone sound:

The richness of the Saxophone sound in Formant-bands of strong intensity in the range of 2.000-8.000Hz is demonstrated in Figure 6. Although the basic frequencies of the played three tones are low

(196Hz; 220Hz; 249Hz) the respective Formant-spectra are very similar and show up to 15 Formant-bands in the range of 1.000-10.000Hz, several with high intensity. (see also Ref. 8).

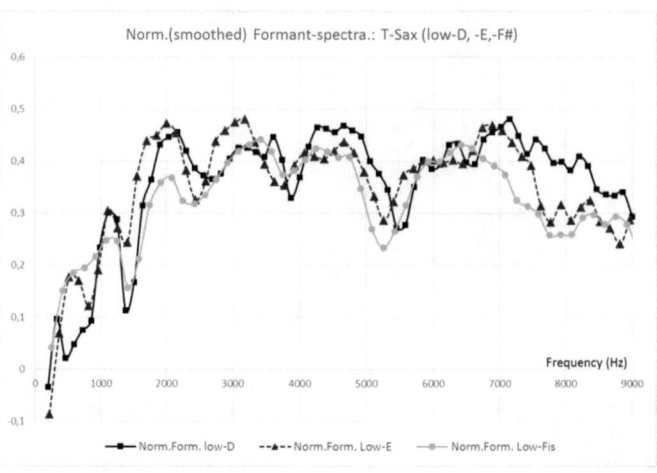

Figure 6: Normalized Formant spectra of three tones played in the lower register of the tenor Saxophone (tones: D, E; F#) by a professional player. Formant spectra a derived from frequency spectra according Ref.8.

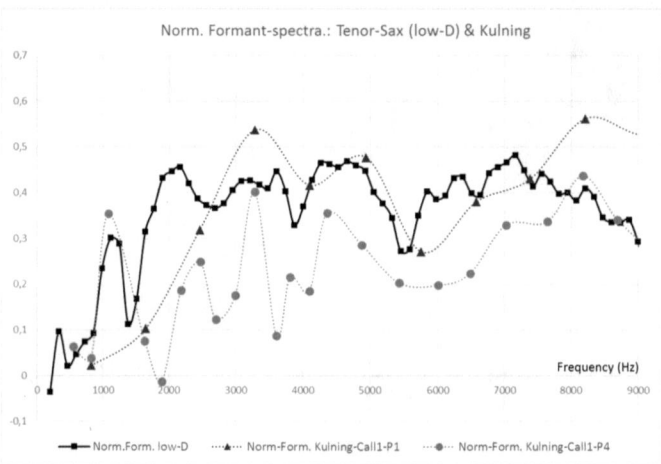

Figure 7: Comparison of the normalized Formant-spectra of Kulning (see Fig. 4) and of the tone "low-D" played on the T-Saxophone (see Fig. 6).

Although the basic frequencies of Kulning (543Hz, 819Hz) and the Saxophone sounds of tones played in the low register (196-249Hz) differ significantly the Formant-spectra of both sounds show high similarities (see Fig. 7). In both cases strong Formant-bands in the range between 1.000-5.000Hz are prominent but also the range of 6.000-9.000Hz is characterized by several Formants with high intensity in both sounds.

Crane call:

The most typical "basic melody pattern of Crane calls" consists of 2 basic and harmonic tones which are varied. Figure 8 shows the frequency spectra of the two harmonic tones of a typical Crane call. In Part 1 of this call the basic frequency is 908Hz, in Part 2 of the call the basic frequency is 1115Hz. It is obvious that the frequency spectra of these Crane calls do not contain strong signals of the overtones in contrast to frequency spectra of Kulning, the Saxophone sound or Cattle calls. Especially in Part 1 of the call there are hardly any overtone signals above 4.000Hz.

Therefore, it is not surprising that the respective Formant-spectra of the two Parts of the Crane call are flat (see Fig 9). Only minor Formant-bands in the range of 2.000-7.500Hz can be observed and in the range >7.500Hz Formant-signals are missing completely.

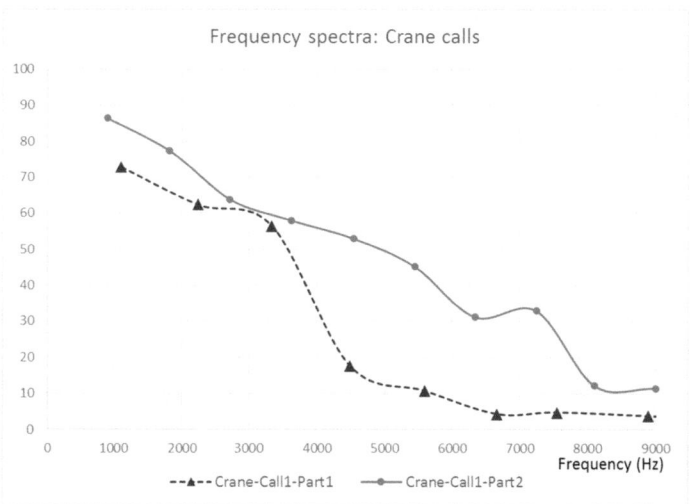

Figure 8: Frequency spectra derived from two parts of a Crane call with stable intensity (dB) and stable basic frequency (Hz). Part 1 has a duration of 300ms with a basic frequency of 1115Hz; Part 2 has a duration of 200ms with a basic frequency of 908Hz. (Y-axis: rel. dB / X-axis: Hz)

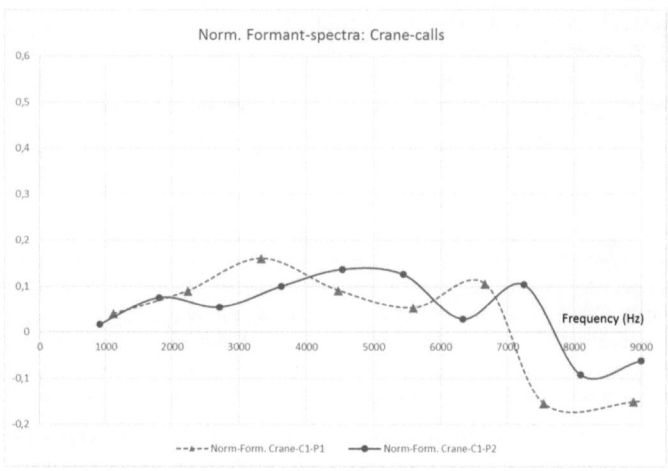

Figure 9: Normalized Formant spectra of the two parts of the Crane call from Fig. 8.

Comparison of Crane call and Saxophone sound:

Comparing normalized Formant-spectra of the important and prominent sound-parts of Crane calls with Formant-spectra of the typical sound of a Tenor-Saxophone (generated played by a professional player) it is obvious that there is a big discrepancy in the intensity of Formant-bands in the range >1.000Hz for both sounds (see Fig.10). In this frequency range Cattle have the best hearing capability as demonstrated in Figure 1.

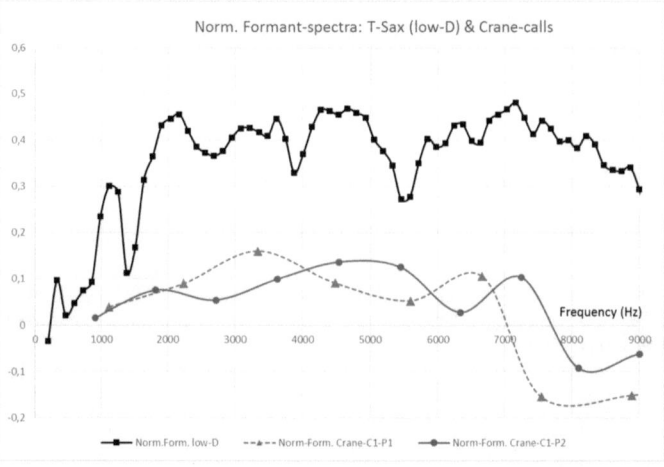

Figure 10: Normalized Formant-spectra of Crane calls from Fig. 9 and of the sound of the low-D played on the tenor-Saxophone from Fig. 6.

8

Discussion:

The attractiveness of a sound for Cattle and its effectiveness to gather Cattle, like this is the case for the ancient herding call "Kulning" (Ref.10) should be judged by taking into account the hearing ability of Cattle in the frequency-range >1.000Hz (Ref. 1; Fig.1) which differs significantly from the hearing ability of humans. As Kulning-calls are obviously very attractive for Cattle and do not contain significant Formant-bands below 1.000Hz it can be assumed that the strong Formant-bands of Kulning in the range >1.000Hz are responsible for this "attractor-effect" on Cattle. Under this assumption it does not surprise that Saxophone sounds, which have a similar Formant-spectrum as Kulning with Formant-signals in the range of 1.000-9.000Hz of high intensity also "attract" Cattle. Although it has been demonstrated that Crane-calls show some variations in dB-intensity in the frequency-range up to 1.200Hz due to Crane individuals (Ref. 11), the Formant spectra of the typical Crane calls show only minor Formant-signals in the range >1.000Hz. This means that Crane-calls are dominated by the basic frequency and so the "timbre" or sound of such calls are less generated by harmonic overtones in the range of 1.000-9.000Hz. In contrast, the sound or "timbre" of the Saxophone is rich in strong Formant-bands in the range of 1.000-10.000Hz (Ref.7), similar to the sound of Kulning (see Fig. 7). Taking into account that the hearing ability of Cattle in the range of 1.000-9.000Hz is at least as good but most likely even better than of humans, it must be assumed that Cattle do recognize this massive and significant difference in sound or "timbre" of Crane calls and Saxophone play. Based on these findings it can be concluded that i) Cattle have a good ability to differentiate the timber of sounds or basic melodies caused by variations in the intensity of overtones in the range >1.000Hz and that ii) sounds which have a Formant-spectrum with signals of high intensity in the range of 1.000-9.000Hz are "more attractive" for Cattle (like the Saxophone sound) than sounds with a flat Formant-spectrum in this frequency-range (like Crane-calls).

References:

1) R.S.Heffner, H.E.Heffner; "Hearing in Large Mammals: Horses (Equus caballus) and Cattle (Bos Taurus)"; Behavioral Neuroscience; 1983 Vol. 97; No.2; pp 299-309

2) see youtube-file: https://www.youtube.com/watch?v=KvtT3UyhibQ

3) see youtube-file: https://www.youtube.com/watch?v=nyV0IULIKuE

4) see youtube-file: https://www.youtube.com/watch?v=OxpliYeTOIs

5) see youtube-file: https://www.youtube.com/watch?v=QGJgP28Yso4

6) see youtube-file: https://www.youtube.com/watch?v=TdheW61w4Co

7) A.Rehm; L.Rehm; „Schallwellenanalyse des Sounds professioneller TenorsaxophonspielerInnen Teil1: Akustische Komponenten der Schallwelle, die vom Spieler generiert und reguliert werden und den Sound bestimmen"; ISBN: 9783668712769; Deutsche Nationalbibliothek; http://dnb.d-nb.de

8) A.Rehm; „Schallwellenanalyse des Sounds professioneller TenorsaxophonspielerInnen Teil2: Methodik zur Bestimmung und Analyse von Formantenspektren und Formantenbändern aus mittels Fourieranalyse errechneten frequenzabhängigen Intensitätsspektren"; ISBN: 9783668777590; Deutsche Nationalbibliothek; http://dnb.d-nb.de

9) R.Eklund, A.MacAllister, F.Pehrson; "An acoustic comparison of voice characteristics in Kulning, head and modal register"; Proceedings of Fonetik 2013;pp 21-24; ISBN:9789175195827

10) M. Tellenbach Uttmann; 2002; „Eine Untersuchung der Teiltonspektren bei Kulning- und Lockruftechniken"; STM-OnlineVol.5; http://musikforskning.se/stmonline/vol5

11) B.Wesseling; „Frequency analysis of Crane calls and their use for individual recognition"; Conference paper European Crane Conference; November 2000; https://www.researgate.net/publication/247214488

YOUR KNOWLEDGE HAS VALUE

- We will publish your bachelor's and
  master's thesis, essays and papers

- Your own eBook and book -
  sold worldwide in all relevant shops

- Earn money with each sale

Upload your text at www.GRIN.com
and publish for free